G000059790

This Journal Belongs to

Published by

Workplace Wonders

Printed in Great Britain
by Amazon

14412068R00068